Some Notes on Game Bounds

by
Jorge-Nuno O. Silva

ISBN: 1-58112-021-4

DISSERTATION.COM

1998

Blank

UNIVERSITY OF CALIFORNIA

BERKELEY

Some Notes on Game Bounds

A thesis submitted in partial satisfaction of the

requirements for the degree Master of Arts

in Mathematics

by

Jorge-Nuno O. Silva

Professor Elwyn R. Berlekamp, Chair
Professor Hendrik W. Lenstra
Professor Eddie Dekel-Tabak

1991

Copyright

by

© copyright 1991
by
Jorge-Nuno O. Silva

ACKNOWLEDGMENT

I am very thankful to Professor E. Berlekamp for suggesting me this material and for some very fruitful discussions.

I thank also JNICT-Programa CIENCIA for their support.

DEDICATION

This work is dedicated to
Laura, Manuel and Raquel.

Table of Contents

Acknowledgment, ii
Dedication, iii
Table of Contents, iv
Abstract, 1
Chapter 1: Basic Definitions, 2
Chapter 2: Birthdays, 4
Chapter 3: Bounds for Numbers, 5
Chapter 4: Bounds for Infinitesimals, 7
Bibliography, 11

Abstract

In this work we establish some game bounds. For each finite birthday, N, we find the smallest positive number and the greatest game born by day N, as well as the smallest and the largest positive infinitesimals. As for each particular birthday we provide the extreme values for those types of games, these results extend those in [1, page 214].

The main references in the theory of combinatorial games are ONAG [1] and WW [2]. We'll use the notation and some fundamental results from WW—mainly from its first six chapters—to establish some bounds to the size of the games.

Chapter 1

Basic Definitions

Given a game $G = \{L \mid R\}$ where L, R are two sets of games we'll write $G = \{G^L \mid G^R\}$, so G^L (G^R) will stand for the typical Left (Right) option of G.

Definition 1 *The sum of two games $G = \{G^L \mid G^R\}$ and $H = \{H^L \mid H^R\}$ is defined by*

$$G + H = \{G^L + H, G + H^L \mid G^R + H, G + H^R\}.$$

Definition 2 *The negative of a game G is*

$$-G = \{-G^R \mid -G^L\}.$$

We write $G - H$ for $G + (-H)$.

Definition 3 *A game in which the second player to move is the winner is a zero game, 0.*

Definition 4 *The partial order is defined by (see ONAG, page 78, for a more formal approach)*

$G > H$ (G is bigger than H) iff $G - H$ is won by Left, whoever starts.

$G = H$ (G is equal to H) iff $G - H$ is a zero game.

$G < H$ (G is smaller than H) iff $G - H$ is won by Right, whoever starts.

CHAPTER 1. BASIC DEFINITIONS

$G \parallel H$ (G is confused with H) iff $G - H$ is won by the first player to move.

The notation $G \geq H$ ($G \leq H$) means that G is bigger (smaller) than or equal to H.

As all the four possible outcomes of the game $G - H$ have been considered in the definitions above we have always exactly one of the relations:

$$G > H \, , \, G = H \, , \, G < H \, , \, G \parallel H .$$

Definition 5 *G is said to be positive (negative) if $G > 0$ ($G < 0$). If $G \parallel 0$ we say that G is fuzzy.*

With the order defined above, all games (modulo equality) form a partially ordered commutative group under addition, 0 being the identity and $-G$ the inverse of G.

Definition 6 *G is said to be in canonical form if it has neither dominated nor reversible options (see ONAG, page 111, and WW, Vol I, pages 62, 64).*

Definition 7 *G is a number if G^L and G^R are numbers and no pair of options of G satisfy $G^L \geq G^R$.*

Chapter 2

Birthdays

Definition 8 *The null game,* { | }, *has birthday zero.*

A game G has birthday $N \in \mathbf{N} = \{1, 2, \cdots\}$ if, when written in canonical form, all its options have birthdays at most $N - 1$ and at least one of them has birthday $N - 1$.

On day one three games are born, namely

$$\{0 \mid \ \} = 1, \ \{ \ \mid 0\} = -1, \ \{0 \mid 0\} = *,$$

and eighteen new games are born on day two:

$$\{0 \mid 1\} = 2^{-1}, \ \{0 \mid -1\}, \ \{0 \mid *\} = \uparrow, \ \{-1 \mid 0\} = -2^{-1},$$

$$\{* \mid 0\} = \downarrow, \ \{1 \mid 0\}, \ \{1 \mid -1\}, \ \{1 \mid \ \} = 2, \ \{1 \mid 0, *\},$$

$$\{ \ \mid -1\} = -2, \ \{1 \mid *\}, \ \{0, * \mid 0, *\} = *2, \ \{1 \mid 1\}, \ \{0 \mid 0, *\} = \downarrow *,$$

$$\{-1 \mid -1\} = -1*, \ \{* \mid -1\}, \ \{0, * \mid 0\} = \uparrow *, \ \{0, * \mid -1\}.$$

Chapter 3

Bounds for Numbers

By inspection it is clear that the greatest game born by day N is N, for $N = 0, 1, 2$. We are about to see that this result is general.

Theorem 1 *For each $N \in \mathbf{N}$ the greatest game born by day N is N.*

Proof: We use induction on N. We saw already that this holds for $N = 0, 1, 2$. Suppose G is a game with birthday $N + 1$. We claim that $G \leq N + 1$. Otherwise Right, going first, can win on $N + 1 - G$. As positive integers in canonical form have no Right options, this means that for some Left option of G, G^L, we have

$$N + 1 - G^L \not> 0,$$

which contradicts the induction hypothesis, since G^L has birthday at most N.

An easy consequence of this result is the fact that if G is born on day N or earlier then $-N \leq G \leq N$.

From the exhaustive list for the first days above we see also that the smallest positive number born on day one is 1, and on day two is 1/2. This extends to all other finite birthdays.

Theorem 2 *For each $N \in \mathbf{N}$ the smallest positive number born on day N is 2^{1-N}.*

Proof: We have seen that this holds for $N = 1, 2$. Let G be a positive number born on day $N + 1$, G in canonical form, and suppose that $G \not\geq 2^{-N}$. Then Right, going first, can win on $G - 2^{-N}$. So, as G is positive and $-2^{-N} = \{-2^{1-N} \mid 0\}$, we have $G^R - 2^{-N} \not> 0$ for some Right option of G, G^R. But this contradicts the induction hypothesis, since G^R has birthday at most N. Notice that we assumed G in canonical form and positive, so G^R must be a positive number also.

Chapter 4

Bounds for Infinitesimals

We now turn our attention to the infinitesimal games. G is said to be an infinitesimal game if it satisfies $-\delta < G < \delta$ for every positive number δ. On day zero no infinitesimals are born. On day one $\{0\,|\,0\} = *$ is born which is not a positive game ($*$ is fuzzy), and $\uparrow$ is the unique positive infinitesimal born on day two. Our next result characterizes the smallest of such games for all finite birthdays.

Theorem 3 *For each natural number $N \geq 2$ the smallest positive game born by day N is* $+_{N-2} = \{0\,|\,\{0\,|\,2-N\}\}$.

Proof: For $N = 2$ this is verified by inspecting all of the 22 games above, since $+_0 =\, \uparrow$. Assume the result holds for all birthdays at most $N - 1$. Let G be born on day N, $G > 0$. Suppose, if you can, that $G \not\geq +_{N-2}$. This implies that Right, going first, can win on

$$G + \{\{N - 2\,|\,0\}\,|\,0\}\,.$$

As G is positive, we must have, for some Right option of G, G^R,

$$G^R + \{\{N - 2\,|\,0\}\,|\,0\} \leq 0\,,$$

which implies that Right, going second, can win on

$$G^R + \{\{N - 2\,|\,0\}\,|\,0\}\,.$$

In particular,

$$G^R + \{N-2\,|\,0\} \not\geq 0.$$

Then, for some Right option of G^R, G^{RR}, we must have

$$G^{RR} + N - 2 \not\geq 0,$$

but this is contradicted by Theorem 1, since G^{RR} was born on or before day $N-2$.

We next show that $N\cdot\uparrow$ and $N\cdot\uparrow\!*$ are maximal among the infinitesimals with birthday at most $N+1$.

We recall that in canonical form we have (see WW, Vol I, page 73)

$$N\cdot\uparrow = \{0\,|\,(N-1)\cdot\uparrow\!*\} \quad \text{and} \quad N\cdot\uparrow\!* = \{0\,|\,(N-1)\cdot\uparrow\}.$$

Theorem 4 *If G is a game with birthday at most $N+1$ ($N \in \mathbf{N}$) then we have at least one of the following:*

$$i)\ G \leq N\cdot\uparrow$$

$$ii)\ G \leq N\cdot\uparrow\!*$$

$$iii)\ L(G) \geq 2^{1-N}$$

$$iv)\ G = 2^{-N}$$

where $L(G)$ stands for the Left stop of G.

Proof: We use induction. For $N = 2$ the result is clear, by inspection. Let G be a counterexample, in canonical form, of minimal birthday, $N+2$. Then we have

$$1.\, G \not\leq (N+1)\cdot\uparrow,\ 2.\, G \not\leq (N+1)\cdot\uparrow\!*,\ 3.\, L(G) < 2^{-N},\ 4.\, G \neq 2^{-1-N}.$$

Hence

CHAPTER 4. BOUNDS FOR INFINITESIMALS

and

$$1.\,a)\,G^L \geq (N+1)\cdot\uparrow \quad \text{or} \quad 1.\,b)\,G \geq N\cdot\uparrow*$$

$$2.\,a)\,G^L \geq (N+1)\cdot\uparrow* \quad \text{or} \quad 2.\,b)\,G \geq N\cdot\uparrow\,.$$

Also from 3. we get $L(G) \leq 0$, since stops are numbers, and, by Theorem 2, there are no smaller positive numbers by this day. So, for every Left option of G, G^L, we have $R(G^L) \leq 0$, where $R(G^L)$ stands for the Right stop of G^L, which implies

$$3.\,a)\,G^L = 0$$

or that some Right option of G^L, G^{LR}, satisfies $L(G^{LR}) \leq 0$. By induction we have in this case

$$3.\,b)\,G^{LR} \leq (N-1)\cdot\uparrow \quad \text{or} \quad 3.\,c)\,G^{LR} \leq (N-1)\cdot\uparrow*\,.$$

2. b) and 3. b) imply

$$G^{LR} \leq (N-1)\cdot\uparrow \leq N\cdot\uparrow \leq G$$

so G^L would be reversible, contradicting the canonical form of G.

From 2. a) we get

$$(N+1)\cdot\uparrow* \not\geq G^{LR}$$

which contradicts 3. b). So 3. b) is impossible.

From 1. a) we have

$$G^{LR} \not\leq (N+1)\cdot\uparrow$$

which contradicts 3. c). And from 3. c) we get

$$G^{LR} \not\geq (N-2)\cdot\uparrow$$

that is inconsistent with 1. b). So we can not have 3. c) either.

CHAPTER 4. BOUNDS FOR INFINITESIMALS

3. a) is clearly incompatible with each of 1. a) and 2. a), so we must have 1. b), 2. b) and 3. a) simultaneously.

From 1. b) and 2. b) we deduce, respectively,

$$G^R \not\leq N \cdot \uparrow * \text{ and } G^R \not\leq N \cdot \uparrow,$$

so, by induction, we have

$$5.\,a)\; L(G^R) \geq 2^{1-N} \quad \text{or} \quad 5.\,b)\; G^R = 2^{-N}.$$

5. a) implies that some Left option of G^R, G^{RL}, satisfies

$$G^{RL} > 2^{1-N} - \delta$$

for every positive number δ, and from 3. a) we get $L(G) = 0$, so $G < G^{RL}$, contradicting the canonical form of G. We conclude then that 5. a) is impossible.

3. a) and 5. b) imply

$$G = \{0 \mid 2^{-N}\} = 2^{-N-1}$$

contradicting 4. Hence no counterexample can exist.

Bibliography

1. Conway, J. H., *On Numbers and Games*, Academic Press, London 1976.

2. Berlekamp, E. R., Conway, J. H., Guy, R. K., *Winning Ways*, Academic Press, London 1985.

Notes

www.ingramcontent.com/pod-product-compliance
Lightning Source LLC
Chambersburg PA
CBHW030855180526
45163CB00004B/1592